PHYSICAL SCIENCE

Concepts and Theory

Dr. Peter Kibet

Description

This e-book book is primarily aimed at high school students and university students who are interested in the structure, properties, dynamics and spectroscopy of atoms, molecules, biological systems, radioactivity, gas laws, and force. This book is for all chemistry, physics or physical science students.

About the author

Dr. Peter Kibet is a lecturer at Machakos University Department of Early Childhood Education where he teaches courses in reading, language arts and developmental psychology for undergraduate, masters and PhD students. Previously, Kibet taught at Kenyatta University, Mt. Kenya University and Ndalat High School. He has 23 years experience teaching in high school and University. He has published widely in referred journals.

Acknowledgement

My heartfelt thanks go to Machakos University students who taught me as I taught them and my son Brian who type parts of this e-book.

TABLE OF CONTENT

Contents

CHAPTER ONE

FORCE

1.0 Introduction

Force is believed to be a pull or a push of objects. It is felt when one is walking, sitting, rubbing things together, when objects are at rest among other activities. The SI unit of force is Newton (N). 1 newton is the same as 0.1kilograms i.e. 1kg = 10N. Force has different effects it can move stationary objects, it can stop moving objects, it can distort the shape of objects and it also changes direction of moving objects. Move is applied in our everyday life.

1.2 Types of force

There are many forms of force which include:

  i.   Gravitational force
  ii.   Magnetic force
  iii. Frictional force
  iv.   Adhesive force
  v.   Cohesive force
  vi.   Centripetal force

a) Gravitational force.

One day newton was lying under a mango tree a mango fell; he questioned himself why it did not go up instead fell down. He said that there must be a force that pulls objects towards the center of the earth. This force he said that it is called gravitational force. Therefore, gravitational force it is a force that is acting towards the center of the earth.

Convert the following kilograms into newton:

  i.   10kg
        If 1kg =10N/kg
        $\cdot$: 10kg=?
        $$\frac{10kg \times 10N/Kg}{1}$$

        =100N

Convert the following into kilograms:

  ii.   1000N
        If 1kg=10N/kg
        1000N= ?
        $\underline{1000 \times 1}$

=100Kg

b) *Magnetic force*

This is a force that is felt when a magnet attracts magnetic materials. A magnet has got two ends called poles i.e. North Pole and South Pole. When a magnet is hung freely on the earth, it will rotate and rest on the direction of North-South direction. A magnet exhibits two types of forces: attraction and repulsion forces. When a north pole is brought close to another north pole of a different magnet they will repel i.e. like poles repel while unlike poles attract. When unlike poles are brought together they attract. This gives us the law of magnetism: like poles repel while unlike poles attract.

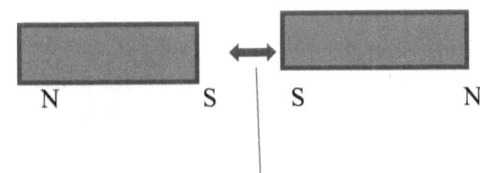

Repulsion within like poles

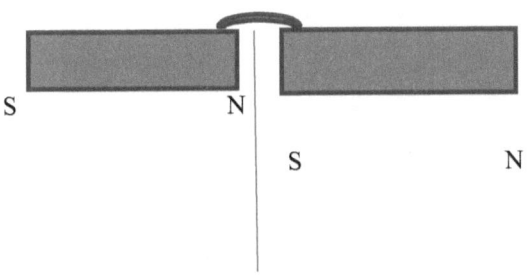

Attraction within unlike poles

c) *Frictional force*

This is a force that opposes motion. It is felt in objects that are moving on one another. The force has both benefits and losses. It is applied in walking without sliding, putting on clothes, rubbing hands together, wear and tear of clothes and soles. The frictional force can be reduced through:

i.   Greasing
ii.   Smoothening rough surfaces

iii.   Application of oil (oiling)

Frictional force = apparent force –real force

E.g. a man uses 500N when pushing a 20kg load on a rough floor. Calculate frictional force that opposed the movement.

Frictional force = apparent force –real force

$$=500N-(20\times10) \text{ N}$$

$$=500-200$$

$$=300N$$

c) *Adhesive and cohesive forces*

Adhesive force is a force felt when there is attraction between unlike molecules. Cohesive force is a force between molecules of different kind. For example water wets a glass when poured on it i.e adhesive force but when wax is smeared on the glass then water is poured on it the water forms balls like drops i.e. there is high cohesive force between the water molecules.

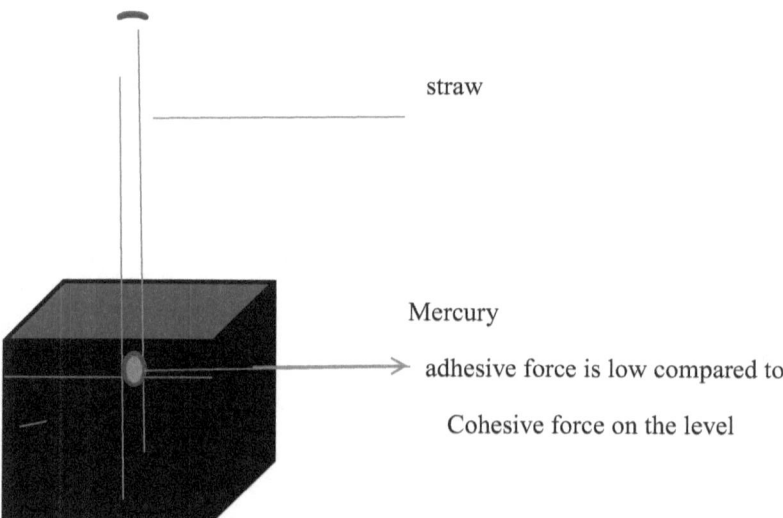

straw

Mercury

adhesive force is low compared to

Cohesive force on the level

In the above diagram it is evident that there is greater cohesive force between the molecules of mercury compared to adhesive force between the straw and the mercury.

d) *Centripetal force*

This is a type of force that is felt when an object negotiates a corner. It is also felt when objects move round a corner. When objects negotiate corners there is a force that pulls them towards the center of the circle in which they negotiate. The force that keeps the objects on track without skidding off is called centripetal force.

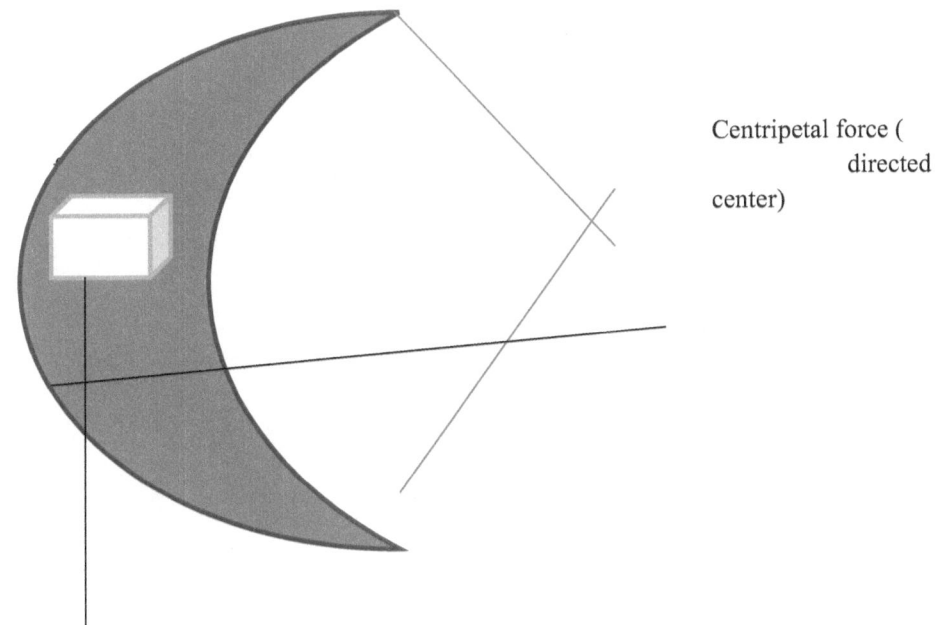

Centripetal force (
directed
center)

Round corner moving object

1.3 Conclusion

Force as it has been discussed above is the pull or a push of objects. These forces are applied in our daily life through walking, writing using pens and pencils, sweeping, putting on clothes sharpening of objects among others. There are other forms of force which include: centrifugal force, electrostatic force, electromagnetic force and frictional force. All this force helps human beings to live comfortable.

CHAPTER TWO

GAS LAW

2.0 Introduction

Matter exist in three forms i.e. liquids solids and gases. In this topic we are going to discuss how pressure and temperature affect the properties of gas as matter. Gas as matter has got properties among them is that gases have indefinite volume, indefinite shape and it has got definite mass. Temperature affects gases in that when the temperature is increased pressure increases on the walls of the container if only volume is kept constant. Likewise, if the volume of a gas is reduced then the pressure increase only when the absolute temperature. Boyle and Charles based their argument on the three factors that is temperature, pressure and volume.

2.1 Boyle`s Laws

It deals with the relationship between pressure and volume of a fixed mass of a gas when temperature is kept constant.

2.1.1 Law

- The law states that: *The volume of a given mass of a gas is inversely proportional to its pressure at constant temperature.*
- The pressure in gases is as a result of collision of the gas molecules with the walls of a container containing the gas.
- In Boyles law the volume of a gas decreases and pressure increases when the mass and temperature is kept constant.

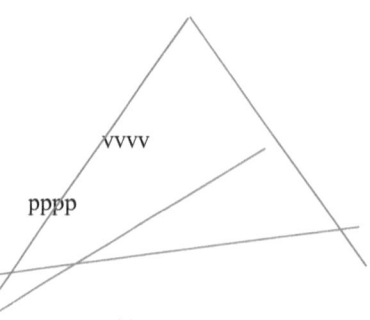

vvvv

pppp

- This means that: $V\alpha 1/p$ where V ~volume, P~pressure

  :'$V\alpha 1/p$

5

V=k/p

PV=K (where k is a constant)

When p_1 changes to P_2 and V_1 changes to V_2

Therefore, $P_1V_1=K$...........i

$$P_2V_2=K...........ii$$

Combining equation i and ii .$P_1V_1=P_2V_2$

- A graph of pressure against 1/v is a straight line graph through the origin.

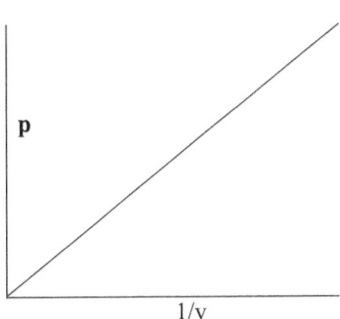

Graph of pressure against 1/v

2.2 Charles` Law

The law deals with the relationship between the volumes of a fixed mass of a gas with its temperature at constant pressure. In the law Charles came up with a conclusion that there is a relationship between temperature and volume when the absolute temperature is kept constant.

Law

The law states that: *The volume of a given mass of a gas is directly proportional to its absolute temperature when its pressure is kept constant.*

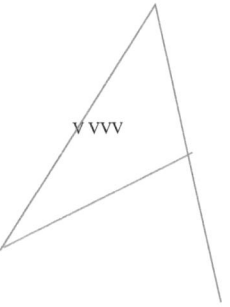

6

This means that $V \alpha T$

i.e $V \alpha T$

Introducing constant K

$V = KT$

$V/T = K$

When V_1 changes to V_2 and T_1 to T_2

Then $V_1/T_1 = V_2/T_2$

Absolute zero- It is the temperature at which the volume of a gas is assumed to be zero.

Ideal gas-It is a gas whose volume is theoretically zero.

2.3 Combined Gas Law

The law is a combination of Charles` law and Boyles` law where the effect of temperature, volume and pressure is considered.

Law

The law deals with the variation in the volume of a fixed mass of a gas with respect to changes in temperature and pressure. The mathematics expression of Charles and Boyle's law is combined.

i.eCharles law $V \alpha T$

Boyles law $V \alpha 1/p$

:` $V \alpha T/P$ Introducing constant K in the expression

$V = KT/P$

$PV = KT$

$PV/T = k$

When P_1 changes to P_2, V_1 to V_2 and T_1 to T_2 then:

$V_1P_1/T_1=V_2P_2/T_2$ (this is an ideal gas equation)

2.4 Applications of Gas Laws
- Inflating tires, balls and balloons depends on the prevailing temperature conditions.
- Designing of aerosol cans and tear- gas canisters.
- Regulations of pressure in an aircraft for comfortable in flight environment at high altitude.

2.5 Conclusion
Gas as matter it occupies space and is affected by the temperature. It is clear that when temperature is increased and the volume is kept constant pressure is increased. This is because the intermolecular space of the air is increased. This is due to increased kinetic energy which in turn exerts pressure on the walls of the containers they are held in. increase in temperature increases kinetic energy.

CHAPTER THREE

RADIOACTIVITY

3.0 Introduction

Radioactivity is a nuclear process which happen the nucleus of the atom. The process involves the protons and neutrons of the atom. During the process a lot of energy is emitted and the atom disintegrates to form new elements. This process is the one that is applicable in the making of the atomic bombs. Uranus for example disintegrates emitting new elements and a lot of energy is produced.

3.1 Radioactivity

Radioactivity is the process where an unstable nuclides breaks up to yield new nuclides of different composition with emission of particles and energy. Radioactive are substances that undergo radioactivity while Radioisotopes are isotopes that are radioactive. Radioactivity is the nuclear process not a chemical process i.e. the nucleus is involved in the reaction not the electrons

Radioactive decay- it is the spontaneous disintegration of radioactive nuclides.

*Isotopes-*these are atoms of the same elements with different mass numbers

3.2 Types of Radioactivity
- *Artificial radioactivity-* occurs when large stable nuclides are bombarded with fast moving energy particles
- *Natural radioactivity-* occurs when radioactive nuclei splits spontaneously yielding a new nuclide with emission of radiations and energy

3.3 Types of Radiations
-alpha (α)

- Beta (β)

- Gamma (γ)

3.4 Characteristics of radiations
- α- these are particles that are positively charged Helium nuclei $_2^4\text{He}^{2+}$
- α- they are the heaviest

9

- β- they are negatively charged electron $_{-1}^{0}e$
- beta particles are formed when a neutron changes into a proton within the nucleus

    $_{0}^{1}n \longrightarrow {}_{1}^{1}p + {}_{-1}^{0}e$
- Gamma (γ) radiations are high energy rays. They have no electrical charge.

3.5 Properties of radiations
1. *The Ionizing power of radiations*
- Alpha has the highest ionizing power because they are the heaviest and moves through the air slowly. This makes them to collide with a lot of the air particles hence high ionizing power. They cause positive ionization on the air around it.
- Beta particles are negatively charged and charge the air around it negative. They are lighter than alpha therefore they move faster in the air charging the air less than alpha. It has a less ionizing power than alpha.
- Gamma radiations have no charge and do not have any effect on the air around it.

2. *Deflection by an electric field*
- αare positively charged and therefore are attracted to the negatively plate of the electric field. They are slightly defected because they are the heaviest.
- β are negatively charged and therefore attracted to the positive plate in the electric field. The defection is greater than of α because they have smaller masses.
- γ negative rays have no charge and are therefore unaffected by an electric field.

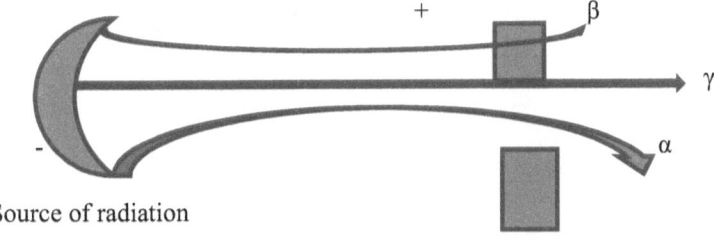

Source of radiation

3. *Ionizing effect of radiation*
- α particles have very high ionizing power ,they produce large numbers of ions where they pass through because of low speed hence high charge.
- B have lower ionizing power and produce fewer ions in gases.

10

- γ have very low ionizing power. This is because of its fast movement as they pass through medium.

4. *Penetrating power*
- α -particle is the heaviest and has the lowest penetrating power. It is therefore blocked by a piece of paper.
- β- Particle moves faster in the air and has a higher penetrating power than alpha particles. It penetrates the piece of paper but is blocked by an aluminum foil.
- Gamma rays are the fastest particles they have a highest penetrating power; it penetrates through the aluminum foil but blocked by a thick block of lead.

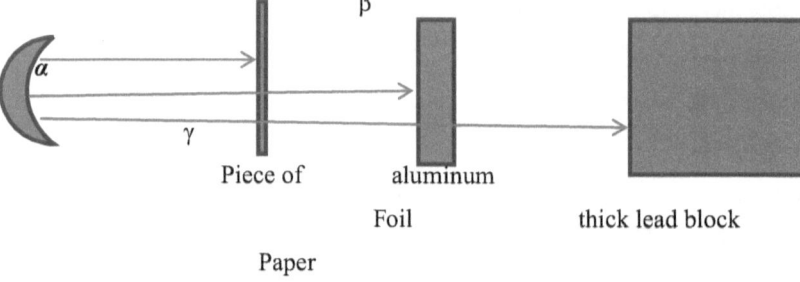

Piece of
Paper

aluminum
Foil

thick lead block

3.6 Radioactive Decay and Half-life
- When a radioactive decays a new nuclide is formed. This is disintegration. As disintegration proceeds fewer and fewer unstable atoms remain and energy is produced. When disintegration occurs the original mass of the nuclides decays to it's a half i.e. a half- life. Half-lifeof a radioactive isotope is the time taken for a given mass or number of nuclides to decay to half its original mass or number. E.g.400-200-100-50
- Or

  This formula is used

  Remaining amount = $(\frac{1}{2})^n$ x original amount.

  ✓ Where n is the number of half-lives e.g. calculate the remaining amount after 24.3 days if the original mass was 400g and the half-life was 8.1 days

$$= (1/2)^3 \times 400$$
$$= 50g$$

3.7 Nuclear Reactions

- If an α particle is emitted both the mass number and atomic number decreases forming a number is covered by ^{2} and mass number by 4.

$$_{91}{}^{23}pa \xrightarrow{\alpha 229} {}_{89}Ac + {}^4_2He$$

- If beta particle is emitted an electron is produced where the atomic number increases by one as the mass number remains. E.g.

$$_6{}^{14}C \xrightarrow{\beta} {}_7{}^{14}N + {}_{-1}{}^0e$$

$$_{88}{}^{226}n \xrightarrow{\beta} {}_{89}{}^{226}Ac + {}_{-1}{}^0e$$

Gamma rays are a form of energy which accompanies other radioactive emissions. They are produced when the remaining particles in the nucleus re-organizes themselves into more stable arrangements. They are not shown when writing nuclear equationsbecause they have no effect on mass number and atomic number. E.g.

$$_{92}{}^{236}u \xrightarrow{\hspace{3cm}} {}_{56}{}^{141}Ba + {}_{36}{}^{92}Kr + 3{}^1_0n + energy$$

3.8 Nuclear Fission

It is the splitting process where a heavy nuclide undergoes when bombarded by fast moving neutron. During the process much energy is liberated e.g.

$$_{92}{}^{236}u \xrightarrow{\hspace{3cm}} {}_{56}{}^{141}Ba + {}_{36}{}^{92}Kr + 3{}^1_0n + energy$$

The energy liberated can be tapped and utilized to generate electrical energy and other forms of energy under controlled conditions. The large heat is used to heat water producing steam which is use to turn turbines to produce electricity.

3.9 Nuclear Fusion

The light nuclear combine together when they are made to collide at high velocity. This fusion results in formation of a heavy nucleus. The process is accompanied by liberation of large quantities of energy.

3.10 Applications of radioactivity

- *Medicine-*where it is used to kill cancerous tissues, sterilizes of surgical instruments using gamma radiation,radioactive iodine is used in patients with defective thyroid to enable doctors to follow the path of through the body, used to monitor growth in bones and healing of fractures.
- *Agriculture* –it is used in monitoring photosynthesis and related process and also monitoring the absorption of phosphate fertilizers.
- *Manufacture of nuclear weapons and atomic bombs*
- *Preservation of food*

3.11 Conclusion

The radiations emitted during radioactivity have got many uses and it improves the lives of people like killing of cancerous cell. Although, radiations emitted have their negative effects on the environment. Longer exposure of the radiations can cause longer term genetic mutation in living tissues leading to anemia, bone cancer and other forms of cancer. The testing of nuclear weapons in the oceans also causes environmental pollution to aquatic organisms. The emission can also be used as weapons of mass destruction as it was experienced in Hiroshima & Nagasaki Japan during the 2nd world war where the effects are felt up to date.

CHAPTER FOUR

BONDING STRUCTURE AND THE PROPERTIES OF MATTER

4.0 Introduction

Chemists use theories of structure and bonding to explain the physical and chemical properties of materials. Analysis of structures shows that atoms can be arranged in a variety of ways, some of which are molecular while others are giant structures. Theories of bonding explain how atoms are held together in these structures. Scientists use this knowledge of structure and bonding to engineer new materials with desirable properties. The properties of these materials may offer new applications in a range of different technologies.

Chemical Bonds

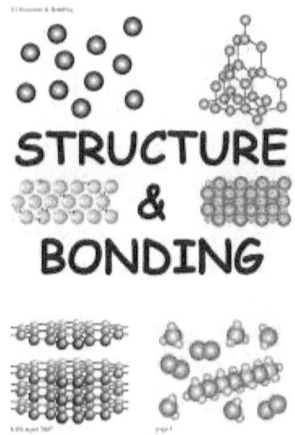

There are three types of strong chemical bonds: *ionic, covalent and metallic*.

- *Ionic bonding* the particles are oppositely charged ions. Ionic bonding occurs in compounds formed from metals combined with non-metals.
- *Covalent bonding* the particles are atoms which share pairs of electrons. Covalent bonding occurs in most non-metallic elements and in compounds of non-metals.
- *Metallic bonding* the particles are atoms which share delocalised electrons. Metallic bonding occurs in metallic elements and alloys.

4.2 Ionic Bonds

When a metal atom reacts with a non-metal atom electrons in the outer shell of the metal atom are transferred. *Metal atoms lose electrons to become positively charged ions. Non-metal atoms gain electrons to become negatively charged ions.*

14

The ions produced by metals in *Groups 1* and *2* and by non-metals in *Groups 6* and *7* have the electronic structure of a *noble gas (Group 0)*. The electron transfer during the formation of an ionic compound can be represented by a dot and cross diagram, eg for sodium chloride.

The *charge* on the *ions* produced by metals in Groups 1 and 2 and by non-metals in Groups 6 and 7 relates to the group number of the element in the periodic table.

4.3 Covalent Bonds

When atoms share pairs of electrons, they form covalent bonds. These bonds between atoms are strong. Covalently bonded substances may consist of small molecules.

Some covalently bonded substances have very large molecules, such as polymers. Some covalently bonded substances have giant covalent structures, such as diamond and silicon dioxide.

4.4 Metallic Bonds

Metals consist of giant structures of atoms arranged in a regular pattern. The electrons in the outer shell of metal atoms are delocalised and so are free to move through the whole structure. The sharing of *delocalised electrons* gives rise to strong metallic bonds

Properties Of Small Molecules

Substances that consist of small molecules are usually gases or liquids that have relatively low melting points and boiling points. These substances have only weak forces between the molecules *(intermolecular forces)*.

It is these intermolecular forces that are overcome, not the covalent bonds, when the substance melts or boils. The intermolecular forces increase with the size of the molecules, so larger molecules have higher melting and boiling points. These substances do not conduct electricity.

4.5 Properties of Ionic Compounds

Properties of Ionic Compounds

Structure:	Crystalline solids
Melting point:	Generally high
Boiling Point:	Generally high
Electrical Conductivity:	Excellent conductors, molten and aqueous
Solubility in water:	Generally soluble

An ionic compound is a giant structure of ions. Ionic compounds are held together by *strong electrostatic forces of attraction between oppositely charged ions*. These forces act in all directions in the lattice and this is called ionic bonding.

Ionic compounds have regular structures (giant ionic lattices) in which there are *strong electrostatic forces* of attraction in all directions between oppositely charged ions.

These compounds have high melting points and high boiling points because of the large amounts of energy needed to break the many strong bonds. When melted or dissolved in water, ionic compounds conduct electricity because the ions are free to move and so charge can flow.

4.6 Giant Covalent Structures

Substances that consist of giant covalent structures are solids with very high melting points. All of the atoms in these structures are linked to other atoms by strong covalent bonds. These bonds must be overcome to melt or boil these substances. *Diamond and graphite* (forms of carbon) and silicon dioxide (silica) are examples of giant covalent structures.

Diamond
In diamond, each carbon atom forms *four covalent bonds* with other carbon atoms in a giant covalent structure, so diamond is very *hard*, has a very *high melting point* and does not conduct electricity.

Graphite

In graphite, each carbon atom forms *three covalent bonds with three other carbon atoms*, forming layers of *hexagonal* rings which have no covalent bonds between the layers. In graphite, one electron from each carbon atom is *delocalised*.

Silicon Dioxide (Silica
Much of the silicon and oxygen in the Earth's crust is present as the compound **silicon dioxide** also known as **silica**. Silicon dioxide has a giant **covalent structure**. Part of this structure is shown in the diagram - oxygen atoms are shown as red, silicon atoms shown as brown:

Each silicon atom is covalently bonded to four oxygen atoms. Each oxygen atom is covalently bonded to two silicon atoms. This means that, overall, the ratio is two oxygen atoms to each silicon atom, giving the formula SiO_2. Silicon dioxide is very **hard**. It has a very **high melting point** (1,610 °C) and **boiling point** (2,230 °C), is insoluble in water, and does not conduct electricity. These properties result from the very strong covalent bonds that hold the silicon and oxygen atoms in the giant covalent structure.

Silicon dioxide is found as quartz in granite, and is the major compound in sandstone. The sand on a beach is made mostly of silicon dioxide.

4.7 Graphene and Fullerenes

Graphene is a single layer of graphite and has properties that make it useful in electronics and composites. Students should be able to explain the properties of graphene in terms of its structure and bonding.

Fullerenes are molecules of carbon atoms with hollow shapes. The structure of fullerenes is based on hexagonal rings of carbon atoms but they may also contain rings with five or seven carbon atoms.

The first fullerene to be discovered was *Buckminsterfullerene (C60)* which has a spherical shape. Carbon nanotubes are cylindrical fullerenes with very high length to diameter ratios. Their properties make them useful for nanotechnology, electronics and materials.

3.8 Polymers

Polymers have very large molecules. The atoms in the polymer molecules are linked to other atoms by strong covalent bonds. The *intermolecular forces* between polymer molecules are relatively strong and so these substances are solids at room temperature..

3.9 Nanoparticles

Nanoscience refers to structures that are 1–100 nm in size, of the order of a few hundred atoms. nanoparticles, are smaller than fine particles (pm2.5), which have diameters between *100 and 2500 nm* (1 x 10-7 m and 2.5 x 10-6 m). coarse particles (pm10) have diameters between 1 x 10-5 m and 2.5 x 10-6 m. coarse particles are often referred to as dust. as the side of cube decreases by a factor of 10 the surface area to volume ratio increases by a factor of 10. nanoparticles may have properties different from those for the same materials in bulk because of their *high surface area to volume ratio.* it may also mean that smaller quantities are needed to be effective than for materials with normal particle sizes

Uses of nanoparticles

Nanoparticles have many applications in medicine, in electronics, in cosmetics and sun creams, as deodorants, and as catalysts. New applications for nanoparticulate materials are an important area of research.

CHAPTER FIVE

ATOMIC STRUCTURE AND PERIODIC TABLE

5.0 Introduction

Atomic structure refers to the structure of atom comprising of a nucleus in which the protons and neutrons are present. The periodic table, also known as the periodic table of elements, is a tabular display of the chemical elements, which are arranged by atomic number, electron configuration, and recurring chemical properties.

5.1 Atomic Structure

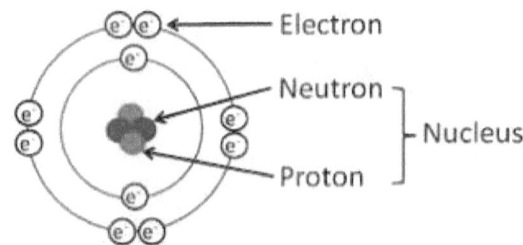

- The particles constituting an atom are the electron, the proton, and the neutron.
- An atom is composed of two regions: the nucleus, which is in the center of the atom and contains protons and neutrons, and the outer region of the atom, which holds its electrons in orbit around the nucleus.
- Protons and neutrons have approximately the same mass, about 1.67 × 10-24 grams, which scientists define as one atomic mass unit (amu) or one Dalton.
- Each electron has a negative charge (-1) equal to the positive charge of a proton (+1).
- Neutrons are uncharged particles found within the nucleus.
- The periodic table is a table that logically organize all the known elements.
- Each row is named "period" where all of the elements have the same number of atomic orbitals.
- Each column is called "group" where the elements have the same number of electrons in the outer orbital
- Every form of matter, or solid, or liquid, or gas, contains atoms.

- Every atom has a basic nucleus at the center, consisting of a certain number of protons and neutrons. The number of this particles are different for different elements. Around the nucleus there is certain number of electrons in fixed orbits, at fixed energy lev
- The electron is by far the smallest: At 9.11×10^{-31} kg. It carries a negative electrical charge. Usually, it is bound to the positively charged nucleus due to the attraction created from the opposite electric charges. If the electrons carried by an atom are more or fewer than its atomic number, then the atom becomes respectively negatively or positively charged. A charged atom is known as an ion.

- Most of the mass of the atom comes from the protons and neutrons themselves, whereas electrons are almost 1/1837th times the weight of a proton or neutron. Protons and neutrons are both composed of other particles called quarks and gluons.
- The atomic number is the number of protons (equal to the number of electrons in a neutral atom) in the atom and the atomic mass number is the sum of the number of protons and neutrons in the atom. The atomic number (Z) is defined as the number of units of positive charges (protons) in the nucleus. It is the number of protons in the nucleus that determines the chemical properties of an atom.
- An atom may gain a positive or negative charge by either losing or gaining electrons respectively. Atoms may attach themselves to each other (of the same type or different type) to form molecules of different compounds, to form matter.
- In some atoms, the nucleus can change naturally. Such an atom is radioactive. In nature, there are some elements that are radioactive, like uranium or radium. In labs, scientists can produce radioactivity by bombarding atoms with smaller particles.
- The molecular mass of a substance is the sum of the atomic masses of all the atoms in a molecule of the substance. It is therefore the relative mass of molecule expressed in atomic mass units (u).
- Isotopes are atoms in a chemical element having different numbers of neutrons than protons and electrons. The atoms in a particular element have the same number of protons and electrons, but can carry varying numbers of neutrons.
- As instance, Hydrogen's atomic number is 1, i.e. its nucleus contains 1 proton. It also has one electron. The Hydrogen atom is neutral since it contains the same number of protons and electrons (as the positive and negative charges cancel each other out).
- However, approximately, one hydrogen atom out of 6000 contains a neutron in its nucleus. These atoms are still Hydrogen because they have one proton and one electron; they simply have a neutron that most hydrogen atoms do not carry. Hence, these atoms are called Isotopes.
- There's also an isotope of hydrogen that contains two neutrons. It's called Tritium, it doesn't occur naturally on earth, but it can easily be created.

5.3 The Periodic Table

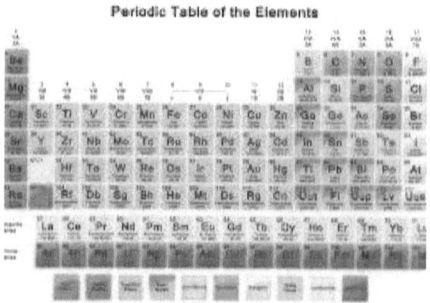

The periodic table is a table that logically organize all the known elements. Each element has a specific location according its atomic structure. Each row and column has specific characteristics. Each row is called a period where all of the elements have the same number of atomic orbitals. For example, every element in the top row (the first period) has one orbital for its electrons. All of the elements in the second row (the second period) have two orbitals for their electrons. As you move down the table, every row adds an orbital. At this time, there is a maximum of seven electron orbitals. Each column is called a group where the elements have the same number of electrons in the outer orbital. Those outer electrons are also called valence electrons. They are the electrons involved in chemical bonds with other elements. Every element in the first column (group one) has one electron in its outer shell. Every element in the second column (group two) has two electrons in the outer shell. Hydrogen (H) and helium (He), in its neutral form, does not have a neutron. There is only one electron and one proton. Helium (He) is very stable with only two electrons in its outer orbital (valence shell). Even though it only has two electrons, it is still grouped with the noble gases that have eight electrons in their outermost orbitals, their valence shell is full. The periodic table can be used also to estimate relatively some other properties of the atoms: electronegativity, ionization energy, electron affinity, atomic radius, melting point, and metallic character. You can use a dynamic periodic table tool to easily refer to specific elements and their atomic number, mass, orbital arrangement, etc.

CHAPTER SIX

ALKANOIC ACIDS (CARBOXYLIC ACIDS)

6.0 Introduction

Alkanoic acids are a class of alkanes which are special with additional of OOH. They are organic acids found mostly in fruits and other vegetable. They have all the characteristics of acids. Some of the acids are found in milk especially sour milk.

6.1 Alkanoic Acids (Carboxylic Acids)

They belong to a homologous series of organic compounds that contains a carboxyl group (COOH) as a functional group which has a formula .They all conform to general formula $C_nH_{2n+1}COOH$ where n is 0,1,2,3………

- Alkanoic acids are naturally found in fruits such as oranges, lemon and pepper.It is also found in nettle leaves and insects stings such as bees and wasps.
- Ethanoic acid is vinegar,Butanoic acid is found in beef fat (butter) ,Hexandioic acid is found in palm oil and olive oil.

6.1.1 Nomenclature

- Alkanoic acids are named by replacing e` in alkanes by the suffix –oic. The simplest member of the alkanoic acid series when n=0 is HCOOH (methanoic acid) and when n=1 is CH_2COOH (ethanoic acid).

6.1.2 Preparation of ethanoic acid

- It is prepared by acidifying potassium manganate (vii) then added ethanol solution to It then heated. On heating the acidified potassium manganate (vii) oxidizes ethanol to ethanoic acid.

$$CH_3CH_2OH_{(aq)} \xrightarrow[\text{MnO4}]{2H^+} CH_2COOH_{(aq)} + H_2O_{(1)}.$$

- During the reaction purple solution of Mn O_4 ions turns to colorless due to oxidation of MnO^{-4} to manganese Mn^{2+} ions. Acidified orange chromate Vi is used after heating it turns green. The chromate (Vi) ions are reduced to green chromium ii ions. Excess oxidizing agent is used to ensure complete oxidation. In order to obtain pure ethanoic acid the mixture is distilled at about $118°c$ where a sharp smell of ethanoic acid is felt.

6.2 Physical properties

i. Alkanoic acids are soluble in water because their molecules are able to form hydrogen bonds with water

ii. Their solubility decreases with increase in their molecular mass because of the decrease in polarity of the acid molecules

iii. Melting point and boiling point increases with increase in molecular mass due to increased van der waal's forces \

iv. The alkanoic acids have higher melting point and boiling point than their corresponding alkanols with the same molecular mass because the alkanoic acids form more hydrogen bond per molecule than the corresponding alkanol

6.3 Chemical properties

i. Alkanoic acids react with metals to form salt and hydrogen gas

   a. $2C_2H_5COOH_{(aq)} + Mg_{(s)} \rightarrow$ $(C_2H_5COO)_2Mg_{(aq)} + H_{2(g)}$

ii. They react with bases to form salt and water only in the process referred to as neutralization

   a. $C_2H_5COOH_{(aq)} + NaOH_{(aq)} \rightarrow$ $C_2H_5COONa_{(aq)} + H_2O_{(l)}$

iii. They react with alkanols in the presence of few drops of concentrated H_2SO_4 liberating a sweet smell in the process called esterification.

$$C_2H_5COOH_{(aq)} + C_2H_5OH_{(aq)} \xrightarrow{H_2SO_4\ (l)} C_2H_5COOC_2H_{5(aq)} + H_2O_{(l)}$$

6.4 Uses of alkanoic acids

i. Used as a solvent in manufacture of drugs and chemicals

ii. In flavoring of foods e.g.ethanoic acids(vinegar)

iii. Manufacture of synthetic fiber such as terylene and nylon

iv. In preservation of food e.g. benzoic acid

6.5 Conclusion

Like other acids, alkanoic acids exhibit similar characteristics as inorganic acids such as hydrochloric acids, sulphuric acids and nitric acids. They are used in the preparation of perfumes and other cosmetics. They are good also in flavoring of food which is good in health.

References

American Association for the Advancement of Science. 1990a. The Liberal Art of Science. Washington, D.C.:   American Association for the Advancement of Science.

American Association for the Advancement of Science. 1990b. Science for All Americans. New York: Oxford University Press.

American Association for the Advancement of Science. 1993. Benchmarks for Science Literacy. New York: Oxford University Press.

American Psychological Association. 1992. Learner-Centered Psychological Principles: Guidelines for School   Redesign and Reform. Washington, D.C.: American Psychological Association.

Anderson, C., and K. Roth. 1992. Teaching for meaningful and self-regulated learning of science. In Advances in Research on Teaching, Vol. 1., J. Brophy, ed. Greenwich, Conn.: JAI.

Angelo, T. A., and K. P. Cross. 1993. Classroom Assessment Techniques: a Handbook for College Teachers,   2nd ed. San Francisco: Jossey-Bass.

Arons, A. B. 1983. Achieving wider scientific literacy. Daedalus Spring: 91-122.

Arons, A. B. 1990. A Guide to Introductory Physics Teaching. New York: John Wiley and Sons.

Astin, A. W., W. S. Korn, and E. R. Riggs. 1993. The American Freshman: National Norms for Fall 1993. Los   Angeles: Higher Education Research Institute, UCLA.

Bailar, J. C. 1993. First-year college chemistry textbooks. J. Chem. Educ. 70:695-698.

Basili, P. A., and P. J. Sanford. 1991. Conceptual change strategies and cooperative group work in chemistry.   J. Res. Sci. Teaching 28(4):293-304.

Baxter-Hastings, N. 1995. Workshop Mathematics: Gateway Courses, access via WWW:http://aug3.augsburg.edu/pkal/aboutpkal.html

Benson, D. L., M. C. Wittrock, and M. E. Baur. 1993. Students' preconceptions on the nature of gases. J. Res. Sci. Teaching 30:587-597.

Berry, D. A. 1987. A Potpourri of Physics Teaching Ideas. College Park, Md. : American Association of   Physics Teachers.

Birk, J. P., and J. Foster. 1993. The importance of lecture in general chemistry course performance. J. Chem. Educ. 70:180-182.Page 78

Suggested Citation:"References." National Research Council. 1997. *Science Teaching Reconsidered: A Handbook*. Washington, DC: The National Academies Press. doi: 10.17226/5287.

Blackburn, T. 1995. From email discussion list posted to cur-l@listserv.ncsu.edu on Feb.15, 1995, subject Scientific misunderstandings, by David Houseman. The archive for this list is located at listserv@ncsu.edu.

Boettcher, J. V., ed. 1993. 101 Success Stories of Information Technology in Higher Education. New York: McGraw-Hill.

Bonwell, C. C., and J. A. Eison. 1991. Active Learning: Creating Excitement in the Classroom. ASHE-ERIC Higher Education Report No.1. Washington, D.C.: The George Washington University, School of Education and Human Development.

Braskamp, L., and J. Ory. 1994. Assessing Faculty Work: Enhancing Individual and Institutional Performance. San Francisco: Jossey-Bass.

Brooks, J. G., and M. G. Brooks. 1993. The Case for Constructivist Class rooms. Alexandria, Va.: Association for Supervision and Curriculum Development.

Brown, D., and J. Clement. 1991. Classroom teaching experiments in mechanics. In Research in Physics Learning: Theoretical Issues and Empirical Studies, R. Duit, F. Goldberg, and H. Niedderer, eds. San Diego, Calif.: San Diego State University.

Caprio, M. W. 1993. Cooperative learning: the crown jewel among motivational-teaching techniques. J. Coll Sci. Teaching 22:279-281.

Cashin, W. E. 1990. Student ratings of teaching: recommendations for use. Idea Paper No.22. Manhattan, Kans.: Center for Faculty Evaluation and Development in Higher Education, Kansas State University.

Cheek, D. W. 1992. Thinking Constructively about Science, Technology, and Society Education. Albany, NY: SUNY Press.

Chickering, A. W., and Z. F. Gamson. 1987. Seven principles for good practice in undergraduate education. Washington, D.C.: American Association of Higher Education. AAHE Bulletin, March:3-7

Claxton, C. S., and P. H. Murrell. 1987. Learning Styles: Implications for Improving Educational Practices. ASHE-ERIC Higher Education Report No.4. Washington, D.C.: Association for the Study of Higher Education.

Clement, J., D. E. Brown, and A. Zietsman. 1989. Not all preconceptions are misconceptions: finding 'anchoring conceptions' for grounding instruction on students' intuitions. Int. J. Sci. Educ. 11:554-565.

Clift, J. C., and B. W. Imrie. 1981. Assessing Students, Appraising Teaching. New York: John Wiley and Sons.

Cooper, M. M. 1995. Cooperative learning: an approach for large-enrollment courses. J. Chem. Educ. 72:162- 164.

Craik, F.M., and R. S. Lockhart, 1972. Levels of processing: a framework for memory research. J. Verbal Learning Verbal Behav. 11:671-684.

Crooks, T. J. 1988. The impact of classroom evaluation practices on students. Rev. Educ. Res. 58(4):438-481.

Davis, B. G. 1993. Tools for Teaching. San Francisco: Jossey-Bass.

Dee-Lucas, D., and J. H. Larkin. 1990. Organization and comprehensibility in scientific proofs, or "Consider a Particle p. . . ." J. Educ. Psychol. 82:701-714.

Dressel, P. L., and D. Marcus. 1982. On Teaching and Learning in College. San Francisco: Jossey-Bass.

Dwyer, F. M. 1972. The effect of overt responses in improving visually programmed science instruction. J. Res. Sci. Teaching 9:47-55.

Ebel, R. L., and D. A. Frisbie. 1986. Essentials of Educational Measurement, 4th ed. Englewood Cliffs, N.J.: Prentice-Hall.Page 79

Eble, K. E. 1988. The Craft of Teaching. San Francisco: Jossey-Bass.

Edgerton, R., P. Hutchings, and K. Quinlan. 1991. The Teaching Portfolio: Capturing the Scholarship in Teaching. Washington, D.C.: American Association of Higher Education.

Erickson, B. L., and D. W. Strommer. 1991. Teaching College Freshmen. San Francisco: Jossey-Bass.

Esiobu, G. O., and K. Soyibo. 1995. Effects of concept and vee mapping under three learning modes on students' cognitive achievement in ecology and genetics. J. Res. Sci. Teaching 32:971-995.

Fraher, R. 1984. Suggestions for beginning teachers. Pp.116-127 in The Art and Craft of Teaching, M. M. Gullette, ed. Cambridge, Mass.: Harvard University.

Fraser, B. J., and K. Tobin. 1989. Student perceptions of psychosocial environment in classrooms of exemplary science teachers. Int. J. Sci. Educ. 11:19-34.

Fraser, B. J. 1986. Classroom Environment. London: Croom Helm.

Freier, G. D., and F. J. Anderson. 1981. A Demonstration Handbook for Physics (2nd ed.). College Park, Md.:   American Association of Physics Teachers.

Fuhrmann, B. S., and A. F. Grasha. 1983. A Practical Handbook for College Teachers. Boston: Little, Brown.

Gabel, D. L., and D. M. Bunce. 1994. Research on problem solving: chemistry. Pp.301-326 in Handbook of   Research on Science Teaching and Learning, D. L. Gabel, ed. New York: MacMillan.

Gardner, H. (1993). Multiple Intelligences: The Theory into Practice. New York: Basic Books.

Gibbons, A. 1993. White men can mentor: help from the majority. Science 262:1130-1134.

Glynn, S. W., and R. Duit. 1995. Constructing Conceptual Models. Pp.1-33 in Learning Science in the   Schools: Research Reforming Practice, S. W. Glynn and R. Duit, eds. Mahwah, N.J.: Lawrence   Erbaum Associates.

Goodsell, A. S., M. R. Maher, V. Tinto, B. L. Smith, and J. MacGregor. 1992. Collaborative Learning: A   Sourcebook for Higher Education. University Park, Pa.: National Center on Postsecondary Teaching,   Learning, and Assessment.

Grosset, J. M. 1991. Patterns of integration, commitment and student characteristics and retention among   younger and older students. Res. Higher Educ. 32(2):159-178.

Hake, R. R. 1992. Socratic pedagogy in the introductory physics lab. Physics Teacher 30:546.

Henes, R. 1994. Creating Gender Equity in Your Teaching. Davis, Calif.: College of Engineering, University   of California, Davis.

Herron, J. D. 1996. The Chemistry Classroom: Formulas for Successful Teaching. Washington, D.C.: American Chemical Society.

Hinton, H. 1993. Reliability and validity of student evaluations: Testing models versus survey research   models. PS: Political Science and Politics September: 562-569.

Holliday, W. G., L. L. Brunner, and E. L. Donais. 1977. Differential cognitive and affective responses to flow   diagrams in science. J. Res. Sci. Teaching 14:129-138.

Hutchings, P.1996. Making Teaching Community Property: A Menu for Peer Collaboration and Peer Review. Washington, D.C.: American Association for Higher Education.

Iona, M. 1987. Why Johnny can't learn physics from textbooks I have known. Am. J. Physics 55:299-307.

Jacobs, L. C., and C. I. Chase. 1992. Developing and Using Tests Effectively: A Guide for Faculty. San   Francisco: Jossey-Bass.

Johnson, D. W., and R. T. Johnson. 1989. Cooperation and Competition: Theory and Research. Edina, Minn.: Interaction Book Co.

Johnson, D. W., R. T. Johnson, and K. A. Smith. 1991. Active Learning: Cooperation in the College Chemistry   Classroom. Edina, Minn.: Interaction Book Co.

Joshi, B. D. 1991. Electronic reports and grading templates for multiple section freshman chemistry   laboratories. J. Comput. Math. Sci. Teaching 10(3):37-49.

Kandel, M. 1989. Grading to motivate desired student performance in a descriptive laboratory course. J. Col. Sci. Teaching 18(4):249-251.

Katz, D. A. 1991. Science demonstrations, experiments, and resources: a reference list for elementary through   college teachers emphasizing chemistry with some physics and life science. J. Chem. Educ. 68(3):235-   244.

Koballa, T. R. 1995. Children's attitudes toward learning science. Pp.59-84 in Learning Science in the Schools: Research Reforming Practice, S. W. Glynn and R. Duit, eds. Mahwah, N.J.: Lawrence Erbaum   Associates.

Kolodny, A. 1991. Colleges must recognize students' cognitive styles and cultural backgrounds. Chronicle   Higher Educ. 37(21):A44.

Kozma, R. B., and J. Johnson. 1991. The technological revolution comes to the classroom. Change 23(1):10-   23.

Lambert, L. M., and S. L. Tice. 1993. Preparing Graduate Students to Teach. Washington, D.C.: American   Association for Higher Education. Laws, P. 1991. Calculus based physics without lectures. Physics   Today Dec:24-31.

Lehman, J. D., C. O. Carter, and J. B. Kahle. 1985. Concept mapping, vee mapping, and achievement: results   of a field study with black high school students. J. Res. Science Teaching 22:663-673.

Lisensky, G., L. Parmentier, and B. Spencer. 1994. Introductory chemistry at Beloit College. Leadership: Challenges for the Future (Occasional paper II). Washington, D.C.: Project Kaleidoscope.

Lowman, J. 1995. Mastering the Technique of Teaching. Second edition. San Francisco: Jossey-Bass.

Marsh, H. W., and M. J. Dunkin. 1992. Students' evaluation of university teaching: a multi-dimensional perspective. In Higher Education: A Handbook of Theory and Research, Vol.8. New York: Agathon Press.

Mayer, M. 1987. Common sense knowledge versus scientific knowledge: the case of pressure, weight and gravity. Pp. 299-310 in Proceedings of the Second International Seminar: Misconceptions and Educational Strategies in Science and mathematics, Vol.1. Ithaca, N.Y.: Cornell University Press.

Mazur, E. 1996. Conceptests. Englewood Cliffs, N.J.: Prentice-Hall.

McDermott, L. C. 1990. A perspective on teacher preparation in physics and other sciences: the need for special science courses for teachers. Am. J Phys. 58(8):734-742.

McDermott, L. C. 1991. What we teach and what is learned—closing the gap. Am. J. Physics 59:301-315.

McDermott, L. C., M. Rosenquist, and E. Van Zee. 1987. Student difficulties in connecting graphs and physics: examples from kinematics. Am. J. Physics 55:503-513.

McDermott, L.C., P. Shaffer, and M. Somers. 1994. Research as a guide for curriculum development: an illustration in the context of the Atwood's machine. Am. J. Phys. 62:46-55.

McDermott, L.C., and P. Shaffer. 1992. Research as a guide for curriculum development: an example from introductory electricity. Part I: Investigation of student understanding. Am. J. Phys. 60:994-1003.

McKeachie, W. J. 1994. Teaching Tips: Strategies, Research, and Theory for College and University Teachers, 9th ed. Lexington, Mass.: D. C. Heath and Company.

Meyers, C., and T. B. Jones. 1993. Promoting Active Learning: Strategies for the College Classroom. San Francisco: Jossey-Bass.

Minstrell, J. 1989. Teaching science for understanding. Pp.129-149 in Toward the Thinking Curriculum: Current Cognitive Research, L. Resnick and L. Klopfer, eds. Alexandria, Va.: Association for Supervision and Curriculum Development.

Moog, R., and J. Farrell, 1996. Chemistry: A Guided Inquiry. New York: John Wiley and Sons.

Moore, J. A. 1984. Science as a way of knowing: evolutionary biology. Am Zoologist 24:421-534.

Murray, H. G. 1991. Effective teaching behaviors in the college classroom. Pg. 135-172 in Higher Education: Handbook of Theory and Research, Vol.7. J. C. Smart, ed. New York: Agathon.

Nakhleh, M. B., and R. C. Mitchell. 1993. Concept learning versus problem solving: There is a difference. J. Chem. Educ. 70(3):190-192.

National Center for Improving Science Education. 1991. The High Stakes of High School Science. Washington, D.C.: National Center for Improving Science Education.

National Research Council. 1996. National Science Education Standards. Washington, D.C.: National Academy Press.

National Science Teachers Association. 1992. The Content Core, Vol. 1. Arlington, Va.: National Science Teachers Association.

Novak, J. D. 1977. A Theory of Education. Ithaca, N.Y.: Cornell University.

Novak, J. D., and D. B. Gowin. 1984. Learning How to Learn. New York: Cambridge University Press.

O'Brien, T. 1991. The Science and Art of Science Demonstrations. J. Chem Ed. 68:933-936.

Okebukola, P. A., and O. J. Jegede. 1988. Cognitive preference and learning mode as determinants of meaningful learning through concept mapping. Sci. Educ. 74:489-500.

Ory, J. C., and K. E. Ryan. 1993. Tips for Improving Testing and Grading. Newbury Park, Calif.: Sage.

Orzechowski, R. F. 1995. Factors to consider before introducing active learning into a large, lecture-based course. J. Coll. Sci. Teaching 24(5):347-349.

Paulos, J. A. 1988. Innumeracy: Mathematical Illiteracy and Its Consequences. New York: Hill and Wang.

Pearsall, M. K., ed. 1992. Scope, Sequence, and Coordination. Vol.11, Relevant Research. Washington, D.C.: National Science Teachers Association.

Pintrich, P. R. 1988. Student learning and college teaching. Pg.61-86 in College Teaching and Learning: Preparing for New Commitments. New

Directions for Teaching and Learning, No.33. R. E. Young and K. E. Eble, eds. San Francisco: Jossey-Bass.

Posner, H. B., and J. A. Markstein. 1994. Cooperative learning in introductory cell and molecular biology. J. Coll. Sci. Teaching 23:231-233.

Posner, G., K. Strike, P. Hewson, and W. Gertzog. 1982. Accommodation of a scientific conception: toward a theory of conceptual change. Sci. Educ. 66(2):211-227.

A Private Universe. 1989. Cambridge, Mass.: Harvard-Smithsonian Center for Astrophysics.

Project Kaleidoscope. 1991. What Works: Building Natural Science Communities. Washington, D.C.: Project Kaleidoscope.

Reynolds, A. 1992. What is competent beginning teaching? A review of the literature. Rev. Educ. Res. 62:1- 35.

Rondini, J. A., and J. A. Feighan. 1978. An ongoing grading technique for laboratory courses. J. Chem. Educ. 55(3):182-183.

Rowe, M. B. 1974. Wait time and rewards as instructional variables, their influence in language, logic, and fate control: Part 1. Wait time. J. Res. Sci. Teaching 11(2):81-94.

Sadker, M., and D. Sadker. 1994. Failing at Fairness: How America's Schools Cheat Girls. New York: Macmillan Publishing Company.

Sandler, B., L. Silverberg, and R. Hall. 1996. The Chilly Classroom Climate: a Guide to Improve the Education of Women. Washington, D.C.: The National Association for Women in Education.

Seymour, E., and N. M. Hewitt. 1994. Talking About Leaving. Factors Contributing to High Attrition Rates Among Science, Mathematics, and Engineering Undergraduate Majors: Final Report to the Alfred P. Sloan Foundation on an Ethnographic Inquiry at Seven Institutions. Boulder, Colo.: University of Colorado.

Shakhashiri, B. Z. 1983. Chemical Demonstrations: A Handbook for Teachers of Chemistry, Volume 1. Madison: University of Wisconsin Press.

Shakhashiri, B. Z. 1985. Chemical Demonstrations: A Handbook for Teachers of Chemistry, Volume 2. Madison: University of Wisconsin Press.

Shakhashiri, B. Z. 1989. Chemical Demonstrations: A Handbook for Teachers of Chemistry, Volume 3. Madison: University of Wisconsin Press.

Shakhashiri, B. Z. 1992. Chemical Demonstrations: A Handbook for Teachers of Chemistry, Volume 4. Madison: University of Wisconsin Press.

Shields, N. 1995. The link between student identity, attributions, and self-esteem among adult, returning students. Sociological Perspectives 38(2):261-272.

Shuell, T. J. 1990. Cognitive conceptions of learning. Rev. Educ. Res. 60(4):531-547.

Shulman, L. 1990. Aristotle had it right: on knowledge and pedagogy (Occasional paper no.4). East Lansing, Mich.: The Holmes Group.

Silberman, M. 1996. Active Learning. Boston: Allyn and Bacon.

Slavin, R., 1989. Research on cooperative learning: consensus and controversy. Educ. Leadership, 47(4):52- 54.

Smith, M. E., C. C. Hinckley, and G. L. Volk. 1991. Cooperative learning in the undergraduate laboratory. J. Chem. Educ. 68:413-415.

Sonnert, G., and G. Holton. 1996. Career Patterns of Women and Men in the Sciences. Am. Sci. 84:63-71.

Stepans, J. 1994. Targeting Students' Misconceptions: Physical Science Activities Using the Conceptual Change Model. Riverview, Fla.: Idea Factory, Inc.

Stover, S. T., G. A. Neubert, and J. C. Lawlor. 1993. Creating interactive environments in the secondary school. Washington, D.C.: National Education Association.

Summerlin, L. R., and J. L. Ealy, Jr. 1985. Chemical Demonstrations: A Sourcebook for Teachers. Washington, D.C.: American Chemical Society.

Summerlin, L. R., C. L. Bogford, and J. B. Ealy. 1987. Chemical Demonstrations: A Sourcebook for Teachers, Volume 2. Washington, D.C.: American Chemical Society.

Tannen, D. 1991. Teachers' classroom strategies should recognize that men and women use language differently. Chron. Higher Educ. 37(40): Bi B4.

Theall, M., and J. Franklin, eds. 1990. Student ratings of instruction: issues for improving practice. New Directions for Teaching and Learning, No. 43. San Francisco: Jossey-Bass.

Thornton, R. K. In press. Microcomputer-based labs and interactive lecture demonstrations. In Proceedings of the Conference on the Introductory Physics Course, J. Wilson, ed. New York: John Wiley and Sons.

Tobias, S. 1978. Overcoming Math Anxiety. New York: W. W. Norton & Company.

Tobias, S. 1990. They're Not Dumb, They're Different: Stalking the Second Tier. Tucson, Ariz.: Research Corporation.

Tobias, S. 1992. Revitalizing Undergraduate Science: Why Some Things Work and Most Don't. Tucson, Ariz.: Research Corporation.

Tobin, K., D. J. Tippins, and A. Gallard. 1994. Research on instructional strategies for teaching science. Pp.45-   93 in Handbook of Research on Science Teaching and Learning, D. L. Gabel, ed. New York: MacMillan.

Trefil, J., and R. M. Hazen. 1995. The Sciences: An Integrated Approach. New York: John Wiley and Sons.

Treisman, U., and R. E. Fullilove. 1990. Mathematics achievement among African American undergraduates   at the University of California, Berkeley: an evaluation of the mathematics workshop program. J   Negro Educ. 59:463-478.

Urbach, F. 1992. Developing a teaching portfolio. Coll. Teaching 41(2):71-74.

Watson, S. B., and J. E. Marshall. 1995. Effects of cooperative incentives and heterogeneous arrangement on   achievement and interaction of cooperative learning groups in a college life science course. J. Res.   Sci. Teaching 32:219-299.

Wendt, D. 1979. An experimental approach to the improvement of the typographic design of textbooks.   Visible Language 13:108-133.

West, I., and A. Pines. 1985. Cognitive Structure and Conceptual Change. New York: Academic Press.

Whimbey, A. 1986. Problem Solving and Cognition. Mahwah, N.J.: Lawrence Erlbaum Associates.

Wilson, J. M. 1994. The CUPLE physics studio. The Physics Teacher. 32:518-523.

Winograd, P., and G. Newell. 1984. Strategic difficulties in summarizing texts. Reading Res. Quarterly   19(4):404-425.

Witkin, H. A., and D. R. Goodenough. 1981. Cognitive Styles: Essence and Origins . New York: International   Universities Press.

Woods, D. R. 1995. Teaching and learning: what can research tell us? J. Coll. Sci. Teaching. 25:229-232.

www.ingramcontent.com/pod-product-compliance
Lightning Source LLC
Chambersburg PA
CBHW030548220526
45463CB00007B/3024